Using Commercial-off-the-Shelf Computer Games to Train and Educate Complexity and Complex Decision-Making

A Monograph
by
MAJ Jeremy R. Lewis
U.S. Army

School of Advanced Military Studies
United States Army Command and General Staff College
Fort Leavenworth, Kansas

AY 2008

SCHOOL OF ADVANCED MILITARY STUDIES

MONOGRAPH APPROVAL

MAJ Jeremy Ryan Lewis

Title of Monograph: Using Commercial-off-the-Shelf Computer Games to Train and Educate Complexity and Complex Decision-Making

This monograph was defended by the degree candidate on 25 September 2008 and approved by the monograph director and reader named below.

Approved by:

_____   Monograph Director
Rob McClary

_____   Monograph Reader
Michael J. Johnson, LTC (P), AR

_____   Director,
Stefan J. Banach, COL, IN   School of Advanced
                                              Military Studies

_____   Director,
Robert F. Baumann, Ph.D.   Graduate Degree
                                              Programs

Abstract

Using Commercial-off-the-Shelf Computer Games to Train and Educate Complexity and Complex Decision-Making by MAJ Jeremy Ryan Lewis, U.S. Army, 46 pages.

Years of U.S. Army involvement in conflicts in Iraq and Afghanistan have underscored the complexity of contemporary operating environments. This complexity within the current operating environment requires U.S. Army leaders not to make simple decisions but complex decisions. Therefore, success on the modern battlefield requires U.S. Army leaders to understand complexity. Current U.S. Army training doctrine does not have a prescribed method for training its leaders in complexity or complex decisions. The Army is relying on its traditional training model of institutional, operational and self-development domains to cover this educational need.

This paper proposes that the U.S. Army use commercial-off-the-shelf (COTS) computer games to train and educate U.S. Army leaders in complexity and complex decision-making. Computer games provide an interactive environment where U.S. Army leaders can experience complex situations and problems and then make complex decisions. Additionally, by immersing U.S. Army leaders into a virtual complex environment, Army leaders have an environment that is free from risk to make decisions. Moreover, timely feedback that computer games provide, allow leaders to see and learn about the intended and unintended consequences of their decisions which creates a learning experience where leaders can innovate and reflect on decisions.

This paper links intellectual capacity – the key leadership attribute that affects decision-making – to complexity and complex decision-making. U.S. Army doctrine breaks intellectual capacity into four characteristics: mental agility, innovation, sound judgment, and domain knowledge. This paper argues that by understanding complexity a leader can make better complex decisions and thereby become more mentally agile, innovative, use sound judgment, and increase domain knowledge. Next this paper presents different classes of computer games, showing how they contain different aspects of complexity and how each game may be used to train and educate U.S. Army leaders in complexity and complex decision-making. Finally, this paper discusses how a COTS computer game training program fits into the Army Training Model and the benefits of using COTS computer games for education and training.

The key finding of this monograph is that the U.S. Army has an opportunity to exploit the benefits of COTS computer games to educate and train leaders. COTS computer games provide a method to educate U.S. Army leaders about complexity and provide a dynamic, immersive environment for practicing the art of making complex decisions.

TABLE OF CONTENTS

Introduction

A New Way to Train

Imagine this scenario. An insurgent hides in palm groves along an irrigation canal. He has a radio transceiver in his hand and is waiting patiently as an American tank approaches. As the tank rumbles cautiously down the road, the insurgent's heart is pounding. The tank rolls into the kill zone, the insurgent presses the detonation button on the transceiver and the improvised explosive device (IED) goes off with an enormous explosion, destroying the tank. The insurgent races off along a previously surveyed escape route and moves into position and prepares for a follow-on attack.

Does this sound like a scenario from Iraq or Afghanistan? Although it may, it is not. It is part of a continuous battle waged in the world of zeros and ones. The difference is that the insurgent sits behind a computer screen, one hand on the keyboard and the other on a mouse. Some would say it is just a game. The insurgent understands that it is more. He understands this game is an opportunity to immerse himself in a virtual environment that arouses emotion, excitement and visual cues that imitate real life. The computer game allows him to practice repeatedly, so when the moment arrives in reality he has already virtually experienced the situations.

Similar to the insurgent, a Duke University Medical Student sits in a classroom monitoring her patient. The patient's heart rate suddenly drops. The student quickly assesses the situation, gives the medical staff a series of commands, and puts her staff in motion. The patient's heart rate returns to normal and the surgery continues as planned. Again, it is easy to assume that this educational process is being conducted in a real operating room dealing with a real patient; however, it is not. Duke University recently began using a first person role-playing game to educate and train anesthesiologists. Students sit behind computers immersed in a role-playing game where they virtually perform the jobs for which they are being educated. The course instructor sits at another console monitoring the situation. The instructor changes the scenario periodically forcing the student to deal with an uncertain, ambiguous, and changing situation. Even though it is not a real operating room, the faculty at Duke University understand the value of training students in an easily changeable environment that provides opportunities for students to communicate with a surgical team in a rapidly changing situation to accomplish a goal.[1]

Both of these examples illustrate the readily available opportunity for an organization to educate and train its members using off-the-shelf technology and software. This technology includes just a few desktop computers, a router, some Local Area Network (LAN) cables and software. Recently, Soldiers and junior leaders within the United States Army also began to utilize this technological opportunity by purchasing commercial off-the-shelf (COTS) computer games with their own money and training on their own time.[2] Many of these Soldiers are using computer games to hone communication skills, teamwork, and problem solving. This growing use of COTS computer games serve as an inexpensive and effective opportunity to train and educate Soldiers in critical leadership tasks.

The U.S. Army's Gaming Future

Last year the U.S. Army established the Training and Doctrine Command (TRADOC) Project Office for Gaming or TPO Gaming. TPO Gaming is currently working to identify whether computer gaming can fill gaps in training that may exist within the U.S. Army. As BG Thomas Maffey, the Director of Training, Army G3, explains, "If Army units are expending

---

[1] Amber Rupinta, "Medical Students Using Games to Practice", WTVD-TV/DT (March 2008) http://abclocal.go.com/wtvd/story?section=news/health&id=5996041, (accessed March 2008).

[2] Michael Peck, "Constructive progress: U.S. Army embraces games — sort of," *TSJ Online: Training & Simulation Journal* (December 2007) http://www.tsjonline.com/story.php?F=3115940, (accessed March 2008).

training funds to purchase games; there is probably an unfilled training requirement."[3] Army leadership, alerted by this development, decided it was time to conduct research into the benefits of computer games as training tools. Specifically, Army leaders want to determine what aspects of COTS computer games are important in developing government off-the-shelf (GOTS) training applications.[4] From TPO Gaming's research, GOTS training applications will be developed and then distributed to the force for training.[5]

Recently, GEN William Wallace, the Commander of TRADOC, asked TPO Gaming, "Does gaming have a leader development application?"[6] While TPO Gaming works to answer this question by develop government-off-the-shelf (GOTS) programs and systems that immerse Soldiers in a virtual world that provides them with an opportunity to train on communication, teamwork and problem solving. This monograph takes a different approach. This monograph is specifically concerned with commercial-off-the-shelf (COTS) computer games and the effectiveness of COTS computer games as an education tool. In short, the research presented herein demonstrates that the U.S. Army should use COTS computer games to train and educate its leaders in complexity and complex decision-making.

This Monograph is organized as follows. First, the main concepts and theory surrounding U.S. Army leadership are defined. Next, this monograph explores complexity and decision-making theory. This understanding of complexity and decision-making provides the motivation for why U.S. Army leaders must be exceptional complex decision-makers. This leads next to recommendations on why and how the U.S. Army should use COTS computer games to educate and train its leaders in complex decision-making. Finally, this monograph concludes that

---

[3] Ibid.

[4] GOTS computer games are government developed computer games and applications versus a commercial product readily available to the public.

[5] Sheldon Parks, meeting with TRADOC Project Office for Gaming, April 2008.

[6] Sheldon Parks, meeting with TRADOC Project Office for Gaming, April 2008.

COTS computer games can be used to immerse leaders in a complex, easily changeable environment where decisions can be monitored and after-action reviews conducted. The naturally dynamic and interdependent nature of COTS computer games combined with their low cost and flexibility mean they are an excellent tool to educate and train leaders in complex decision-making.

Theory and Concepts

The Army has learned, after years of conflict in Iraq and Afghanistan, the current and future operating environments are very complex. This complexity requires U.S. Army leaders to make correspondingly complex decisions. Success on the modern battlefield requires U.S. Army leaders to know how to construct and understand complex system models. These models can be constructed either physically (drawn on a board or computer) or mentally. These models help leaders learn and understand a system and the complexities involved within that system. Once constructed, leaders use these models to create a working simulation, which allows them to test these ideas for managing the complex problem. Therefore, the U.S. Army must develop training programs, which educate and train U.S. Army leaders in complexity and complex decision-making.

This section discusses U.S. Army leadership, decision-making, complexity and complex decisions by identifying the attributes and components of U.S. Army leadership that play an important role for making decisions in a complex environment. Then by focusing on decision-making as outlined by the U.S. Army in its leadership and mission command manuals, several different decision-making models are examined. The third part of this section provides a synopsis of complexity. Finally, complex decisions are defined and a discussion on how they are different from other decision-making models is included.

U.S. Army Leadership

U.S. Army leadership is discussed and defined in the U.S. Army Leadership Field Manual 6-22 (FM 6-22). In FM 6-22, the U.S. Army defines leadership as "the process of influencing people by providing purpose direction and motivation while operating to accomplish the mission in improving the organization."[7] Leaders, according to FM 6-22, are "anyone who by virtue of assumed role or assigned responsibility inspires and influences people to accomplish organizational goals. U.S. Army leaders motivate people both inside and outside the chain of command to pursue actions, focus thinking, and shape decisions for the greater good of the organization."[8] These definitions focus heavily on influencing and motivating individuals or organizations and, in many cases, this is what some believe comprise leadership. Though influencing and motivating are very important aspects of leadership, in today's complex environment, "shaping decisions" may be of more importance. In fact, leader do more than shape decisions, they make decisions.. Therefore, it is imperative that U.S. Army leaders are able to make relevant decisions that produce results, improve the organization and environment in which that organization operates.

---

[7] U.S. Department of the Army, *Army Leadership: Competent, Confident, and Agile: Field Manual 6-22*, (Washington, DC, 2006), 1-2.

[8] Ibid., 1-1.

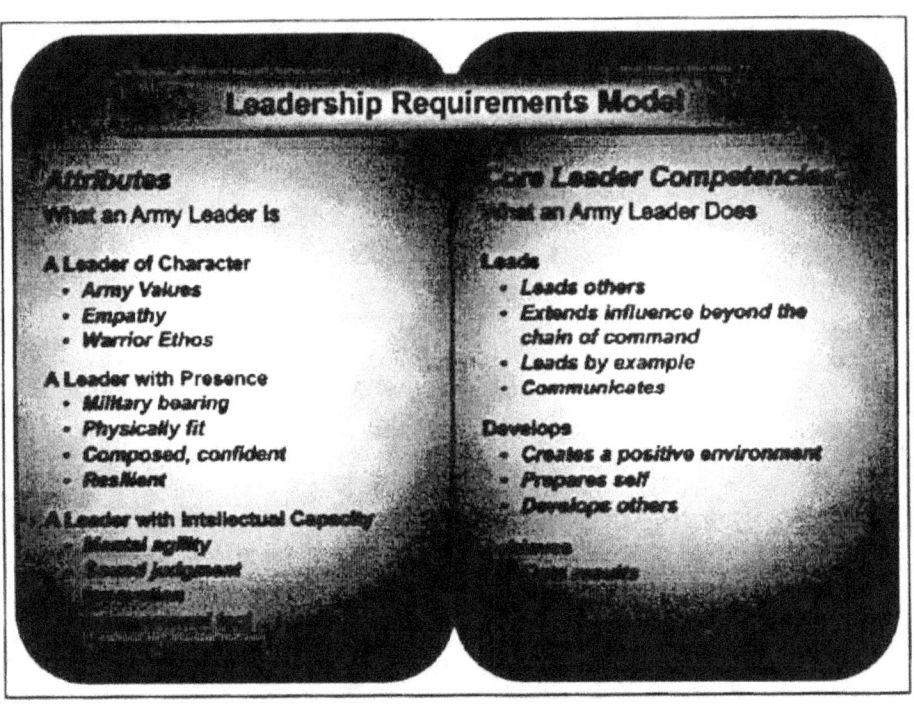

Figure 2.1: U.S. Army Leadership Requirement Model

In FM 6-22, the U.S. Army uses the leadership requirements model (Figure 2.1[9]) to learn and think about leadership.[10] This model breaks leadership into "what an Army leader is" and "what an Army leader does." The attributes for what an Army leader is are character, presence, and intellectual capacity. In the decision-making process, character and presence certainly play a role, but a leader's intellectual capacity greatly affects the decisions made by a leader. Since intellectual capacity is an important attribute in decision-making, this paper focuses primarily on the leadership attribute of intellectual capacity.

FM 6-22 breaks the attribute of intellectual capacity into five components. Those components are mental agility, sound judgment, innovation, interpersonal tact, and domain knowledge.[11] The leadership manual states that mental agility requires a flexible mind, the ability

---

[9] U.S. Department of the Army, *Army Leadership,* 2-4.

[10] Ibid., 2-4.

[11] Ibid., 6-1.

to adapt to a situation or enemy. Mental agility helps leaders to quickly identify and distinguish between simple, complicated, and complex problems.[12] In addition, leaders must understand the consequences (intended and unintended) of their decisions. As FM 6-22 points out, a U.S. Army leader must be a critical thinker– a thinker who tries to understand possible second- and third-order effects of a decision. A critical thinker uses his or her mental agility to understand that any situation is in constant flux and cannot be solved using simple solutions. A mentally agile leader understands that a complex problem requires a complex solution.[13]

Sound judgment comes through experience and development and "requires having a capacity to assess a situation or circumstance shrewdly and to draw feasible conclusions."[14] FM 6-22 poses that sound judgment comes from both good and bad experience. In a sense, leaders must be placed in situations where they can make decisions and learn from those decisions. Unfortunately, sometimes leaders must learn from bad or poor decisions. However, there are ways for leaders to experiment and learn without suffering the actual consequences of poor decision-making, such as field exercises and classroom instructions.

The U.S. Army states that innovation is a leader's ability to "introduce something new for the first time when needed or an opportunity exists."[15] Dietrich Dörner in his book *The Logic of Failure*, describes how decision makers tend to economize their decisions and use preexisting solutions or templates.[16] Economizing balances out a decision maker's slowness to perceive and process information. One technique for economizing is through "methodism." Methodism is

---

[12] Glouberman, S., and Brenda Zimmerman. *Complicated and Complex Systems What Would Successful Reform of Medicare Look Like?* (Commission on the Future of Health Care in Canada, 2002), 2.

[13] William Ross Ashby in his book *Introduction to Cybernetics (1954)* introduces the Law of Requisite Variety. In summation, the Law of Requisite Variety implies that only variety can destroy variety, or as the founder of Management Cybernetics Stafford Beer phrased it, variety absorbs variety. A complex problem is comprised of many variables, to solve a complex problem a solution with as much or a greater variety is required.

[14] U.S. Department of the Army, *Army Leadership*, 6-2.

[15] Ibid.,, 6-2.

looking at a new problem through an old lens. As Dörner points out, it is much easier to use an old solution versus analyzing the situation and realizing it needs a new set of solutions.[17] Therefore, Dörner is critical of methodism, because this type of intellectual shortcut can have serious consequences. One example is the application of strategies found to be successful in Iraq to the conflict in Afghanistan. Although each operating environment is clearly different and requires its own solution and template, some military members seek to use exact techniques from Iraq in Afghanistan. Echoing Dörner, the U.S. Army wants its leaders not to "methodize" but to recognize that new problems may need new solutions. In short, the U.S. Army wants it leaders to be innovative and have the capacity to create new solutions for our world's complex problems.

In any situation, decision-making requires a basic level of knowledge within that field. A leader's knowledge of every aspect in which a leader may operate within his profession is the leader's domain knowledge.[18] FM 6-22 breaks down domain knowledge into technical, tactical, joint, cultural and geopolitical knowledge.[19] This paper takes a constructivist approach and says that a leader, through training, education and experience increases their domain knowledge.[20] A leader builds or constructs their understanding of the environment, their culture and profession by deriving meaning from their experiences. These mental constructs affect how a leader may make a decision and therefore a leader's decisions rely on their domain knowledge. Domain knowledge is a key attribute for making decisions in a complex world.

This section asserted leaders are decision makers. As decision makers, they must have the intellectual capacity to make relevant decisions to improve their organization and

---

[16] Dietrich Dörner, *The Logic of Failure*, (New York, NY: Metropolitan Books, 1996), 187

[17] Ibid.

[18] U.S. Department of the Army, *Army Leadership*, 6-5.

[19] Ibid., 6-5.

[20] Edith Ackerman, *Piaget's Constructivism, Papert's Constructivism: What's the* Difference? (Massachusetts Institute Of Technology), 3.

environment. The U.S. Army states that mental agility, sound judgment, innovation, and domain knowledge are relevant attributes that make up intellectual capacity. Therefore, if the U.S. Army wishes to improve the decision-making skills of its leaders it must then find ways to improve those relevant attributes.

A World of Problems

In the U.S. Army, leaders are paid to recognize, confront and solve problems. Some are simple while others are more complicated and/or complex. For some problems, leaders use entire staffs to think about a problem and formulate a solution or solution set. Other times, leaders must make decisions quickly as they execute a mission under fire. In the process of making a decision, leaders first identify the problem (how the world is different from what he or she would like it to be) and then generate a solution (how they can affect the world to make it, as he or she would like it to be). Problems may be classified based on the amount of understanding, uncertainty and interdependency that exists in the information surrounding the problem. As well as how simple, complicated and complex they are.[21] Understanding the type of problem helps leaders determine what type of decision process is best for solving it.

Most individuals easily understand and solve a simple problem. For example, a Soldier may notice he has a flat tire on a vehicle. The information is readily available as the Soldier can clearly see the flat tire and the solution requires minimal learning. However, in another scenario the vehicle may not start. In this case, information to solve the problem may not be readily available, as there may be a multitude of problems. The Soldier can gather information and if the Soldier is not a mechanic, find a mechanic that is able to identify the problem. In this case, the problem requires someone with specialized training and education to solve this complicated

---

[21] Glouberman, S., and Brenda Zimmerman. *Complicated and Complex Systems What Would Successful Reform of Medicare Look Like?*, 1.

problem. Complicated problems are highly technical and require learning and specialization.[22] Complicated problems are typically deterministic. Given the right specialization and tool sets, an individual may establish the specific causes and formulate a solution.

The final type of problem is a complex problem. A complex problem requires learning, understanding, has interaction among many variables, and is dynamic.[23] Complex problems have a high degree of uncertainty.[24] In the example of the Soldier and the disabled vehicle, a complex problem may encompass the entire system. Maybe the Soldier has broken down along a major road in Iraq. The vehicle needs multiple parts to fix and is immoveable. This road is known for insurgent attacks and there are many locals shopping within the immediate vicinity. The leader of the convoy wishes to move his unit out of the area but does not wish to leave the vehicle behind due to the sensitive nature of equipment aboard. To further the problem, a prominent mosque was recently destroyed and the populace is in unrest and locals are beginning to surround the convoy. All of the locals appear to be unarmed but are definitely agitated. Another facet to the problem, is the recent change in the rules of engagement have made warning shots off-limits as part of a new strategy to reduce the number of shots fired at Iraqi civilians. In this example, the leader faces a complex problem with many variables, and those variables interact at different levels. What appeared to be a complicated malfunction has turned into a situation that could have strategic implications. Additionally, the system is most likely fluid, as different people are involved; some people try to help this stranded convoy, while others may prepare to attack it. The different agents pursue different strategies and interact at different scopes and scales to make for a complex problem. This tyepe of problem requires a complex solution. Once a solution set

---

[22] Glouberman, S., and Brenda Zimmerman. *Complicated and Complex Systems What Would Successful Reform of Medicare Look Like?*, 2.

[23] Ibid.

[24] Hassan Qudrat-Ullah, Michael J. Spector, and Paal Davidsen, *Complex Decision-Making: Theory and Practice,* (Springer, 2008), 1.

is determined adequate for managing a complex problem, a leader must make a decision to put that solution set into action. In this complex situation, the leader must assess the conditions, learn about the system, come to an understanding of the system and environment, and then make a decision.

The U.S. Army requires leaders to make decisions, many of which are not simple or complicated, but complex. To make those complex decisions, the U.S. Army relies on a leader's intellectual capacity. Therefore, if the U.S. Army wishes for its leaders to be proficient decision makers, their leaders must have the intellectual capacity to do so. Therefore, they must be mentally agile, use sound judgment, be innovative, and have superior domain knowledge. By educating its leaders in the art of complex decision-making, the U.S. Army can increase one or all of these attributes, and therefore improve a leader's ability to make better decisions in a complex environment.

Decision-Making

If the U.S. Army should use COTS computer games to educate and train complex decision-making it is important to understand how it thinks about decision-making. Therefore, this section of the monograph discusses two decision-making models used by the U.S. Army, rational decision-making and recognition-primed decision-making (RPD). [25] These models explain how individuals or groups in different situations and circumstances make decisions.

---

[25] Gary A. Klein, *Sources of Power: How People Make Decisions*, (Cambridge, Massachusetts: MIT University Press, 1998), 24.

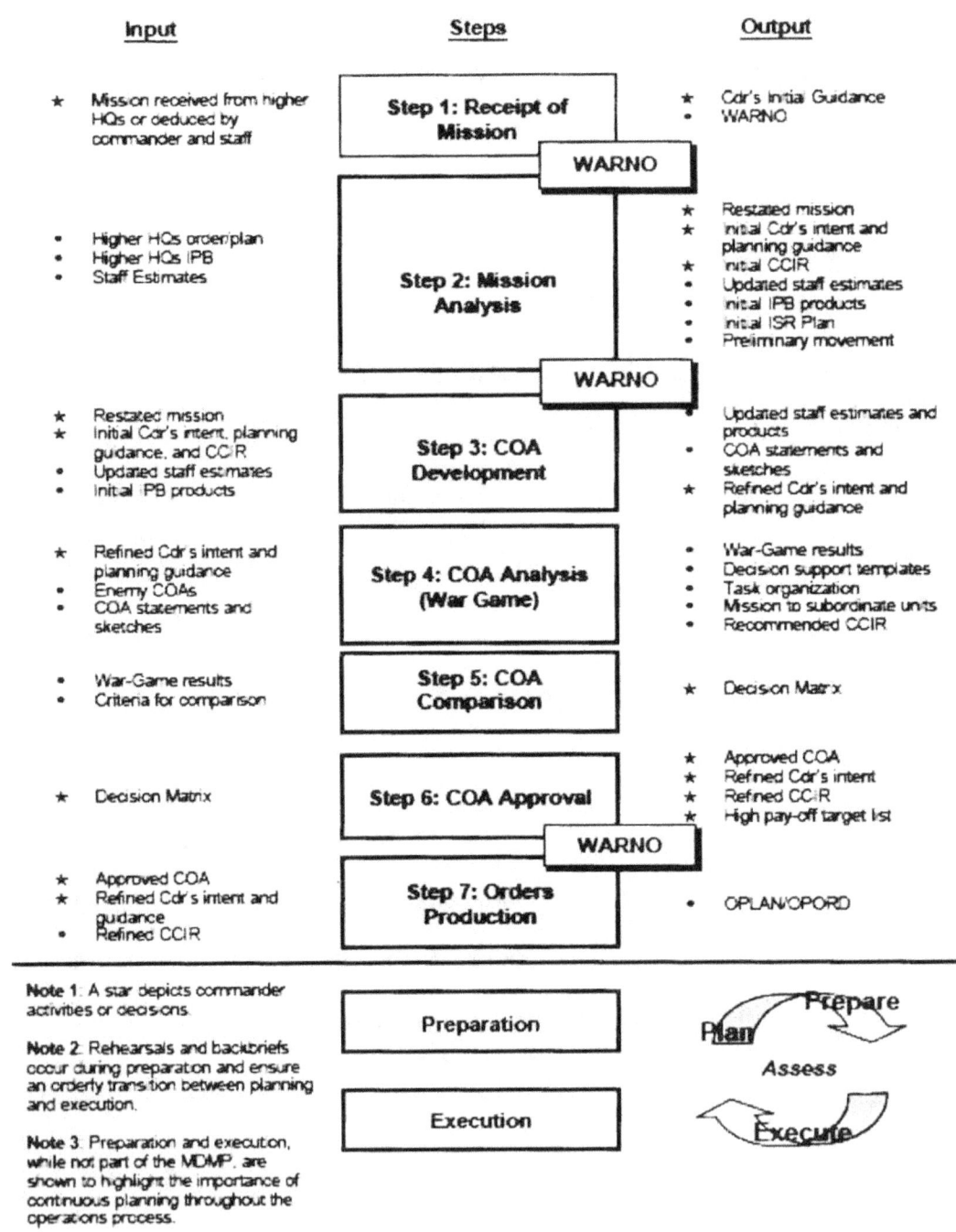

Input	Steps	Output
★ Mission received from higher HQs or deduced by commander and staff	**Step 1: Receipt of Mission**	★ Cdr's Initial Guidance • WARNO
	WARNO	
• Higher HQs order/plan • Higher HQs IPB • Staff Estimates	**Step 2: Mission Analysis**	★ Restated mission ★ Initial Cdr's intent and planning guidance ★ Initial CCIR • Updated staff estimates • Initial IPB products • Initial ISR Plan • Preliminary movement
	WARNO	
★ Restated mission ★ Initial Cdr's intent, planning guidance, and CCIR • Updated staff estimates • Initial IPB products	**Step 3: COA Development**	• Updated staff estimates and products • COA statements and sketches ★ Refined Cdr's intent and planning guidance
★ Refined Cdr's intent and planning guidance • Enemy COAs • COA statements and sketches	**Step 4: COA Analysis (War Game)**	• War-Game results • Decision support templates • Task organization • Mission to subordinate units • Recommended CCIR
• War-Game results • Criteria for comparison	**Step 5: COA Comparison**	★ Decision Matrix
★ Decision Matrix	**Step 6: COA Approval**	★ Approved COA ★ Refined Cdr's intent ★ Refined CCIR ★ High pay-off target list
	WARNO	
★ Approved COA ★ Refined Cdr's intent and guidance • Refined CCIR	**Step 7: Orders Production**	• OPLAN/OPORD

Note 1: A star depicts commander activities or decisions.

Note 2: Rehearsals and backbriefs occur during preparation and ensure an orderly transition between planning and execution.

Note 3: Preparation and execution, while not part of the MDMP, are shown to highlight the importance of continuous planning throughout the operations process.

Preparation

Execution

Plan — Prepare — Assess — Execute

Figure 2.2: The Military Decision Making Process

12

The rational decision-making model is a logical stepped process for making a decision.[26] The U.S. Army Military Decision-Making Process (MDMP) (Figure 2.2[27]) is a type of rational decision-making model and provides a formalized approach to solving problems.[28] The MDMP provides a method for commanders to utilize their staffs to analyze a problem and optimize everyone to explore feasible and suitable courses of action. It is the standard process for decision-making within the U.S. Army in the orders planning and production process[29]. As a stepped analytical decision tool, the MDMP typically requires more time than the other method of intuitive decision-making outlined in the U.S. Army's planning manual.[30]

Another decision-making model is one introduced by Gary Klein. Klein introduced the recognition-primed decision (RPD) model in his book *Sources of Power: How People Make Decision.*[31] After studying firefighters in crises, Klein discovered that in many cases individuals do not develop and consider multiple courses of action for comparison in making decisions. He discovered that people instead develop one course of action based on previous experiences or "people use their experience to make decisions in a field setting."[32] Klein calls this RPD. From these studies, Klein determined that firefighters under extreme time pressures use their experience to identify similarities between the current situation and a previous experience. Next, a mental comparison occurs and the firefighter develops a mental simulation, which they use to reach a decision. Dr. Klein describes the RPD model as a "fusing of two processes: the way decision

---

[26] Stephen P. Robbins and Timothy A. Judge. Organization Behavior. 12th ed. Upper Saddle River, New Jersey: Pearson Prentice Hall, 2007, 156-158.

[27] U.S. Department of the Army, *Army Planning and Orders Production, Field Manual 5-0,* Authpub-FM. Washington, DC, 2006, 3-3.

[28] Ibid., 2-1.

[29] U.S. Department of the Army, *Army Planning and Orders Production,* 2-2.

[30] Ibid., 1-6.

[31] Gary A. Klein, *Sources of Power,* 24.

[32] Ibid., 1.

makers size up the situation to recognize which course of action makes sense, and the way they evaluate that course of action by imagining it."[33] The RPD model is the basis for what the Army calls, intuitive decision-making. FM 5-0 states that intuitive decision-making utilizes a leader's ability to recognize patterns based upon their experience (as discussed by Klein), judgment and intelligence.[34] In FM 5-0, the U.S. Army recommends that experienced leaders use intuitive decision-making in time-constrained environments or during execution when immediate decisions are required.[35] In addition, FM 5-0 points out that intuitive decision-making is not suited for situations where leaders lack experience, there are competing courses of action and the situation is very complex.[36]

These two decision-making models are the main methods outlined within U.S. Army doctrine for making decisions and show how a leader makes a decision. In addition to knowing how a leader makes a decision, it is important to understand the type of decision required for a situation. For example, people make simple decisions all the time; decisions to read a book, to get something to drink or to go to bed. However, some decisions are not simple but complex. Complex decisions are made in a dynamic environment.[37] They may not be single decisions but multiple decisions or decision sets.[38] In addition, they may be interdependent[39] or in competition.[40] The outcomes of complex decisions may be opaque, meaning that causal relationships are not easily recognized or variables affecting the problem are unknown.[41] They

---

[33] Gary A. Klein, *Sources of Power*, 24.

[34] U.S. Department of the Army, *Army Planning and Orders Production*, 1-6.

[35] Ibid.

[36] Ibid., 1-7.

[37] Stefan Strohschneider, *Cultural factors in complex decision-making*, Unit 4, Chapter 1.

[38] Hassan Qudrat-Ullah, Michael J. Spector, and Paal Davidsen, *Complex Decision-Making*, 1.

[39] Ibid.

[40] Stefan Strohschneider, *Cultural factors in complex decision-making*, Unit 4, Chapter 1.

[41] Ibid.

also concern a complex adaptive system. Warfare is a complex adaptive system and therefore, it is imperative that our leaders are able to make relevant complex decisions.

Complexity

For a leader to make relevant complex decisions, they must also understand complexity.[42] COTS computer games provide a virtual environment to practice making decisions, specifically complex decisions.[43] In addition, COTS computer games are full of examples of complexity. These examples, properly presented, provide a rich environment for educating complexity to leaders. Therefore, the U.S. Army should use COTS computer games to train and educate complex decision-making. This section discusses complexity and provides examples of complexity within computer games.

In their book *Harnessing Complexity: Organizational Implications of a Scientific Frontier*, Robert Axelrod and Michael Cohen discuss complex adaptive systems and say a system is complex "when there are strong interactions among its elements, so that current events heavily influenced the probabilities of many kinds of later events."[44] Axelrod and Cohen say a system is adaptive when agents or populations seek to improve according to some measures of success.[45] They define components of complexity to include agents, strategies, variation, interaction, and selection. Not only do computer games contain these components of complexity, but also a teacher can exaggerate the relationships between these components to make them more evident.[46]

---

[42] Stefan Strohschneider, *Cultural factors in complex decision-making*, Unit 4, Chapter 1.

[43] Hassan Qudrat-Ullah, Michael J. Spector, and Paal Davidsen, *Complex Decision-Making*, 1.

[44] Robert Axelrod and Michael D. Cohen, *Harnessing Complexity: Organizational Implications of a Scientific Frontier* (Basic Books, 2000), 7.

[45] Ibid.

[46] Hassan Qudrat-Ullah, Michael J. Spector, and Paal Davidsen, *Complex Decision Making*, 1.

For example, an instructor can accelerate or compress time and space within a computer game to show the effects of player's decisions.[47]

Axelrod and Cohen described agents as something that has the ability to interact with its environment. Importantly, agents have a number of properties that include a location, capability, and a memory.[48] Location of an agent is where an agent operates which can be physical or virtual. Location is where an agent interacts with its environment or other agents while capability is how an agent reacts with its environment or other agents.[49] Agents within computer games can be real players or artificial intelligence computer players. Each of these agents formulate strategies to accomplish goals in order to win or progress in the game. Some games may be single player with just one human agent, while other games have many human agents. In either case, the agents or players must choose their strategies in order to progress.

An agent's memory is what it carries forward from its past that allows it to remember favorable strategies. Axelrod and Cohen state that strategy is "The way an agent responds to its environment and pursues its goals."[50] Strategies can be deliberate choices but can also be responses that pursue goals with little deliberation. The strategy an agent selects is what helps an agent adapt within a complex system. In computer games, strategies are continuously evolving. For example, some players learn by observing successful strategies of others, and players also share their successful strategies with friends and publicly on gaming forums.. Additionally, a strategy adapts as new players add variety and innovation to imperfectly copied strategies.

An important aspect in adaptation and emergence of new strategies is variation. Variation provides the means for adaptation to occur. Without variation, adaptation is slow or

---

[47] Hassan Qudrat-Ullah, Michael J. Spector, and Paal Davidsen, *Complex Decision Making*, 1.

[48] Robert Axelrod and Michael D. Cohen, *Harnessing Complexity*, 4.

[49] Ibid., 6.

[50] Ibid., 4.

nonexistent. Axelrod and Cohen describe variation as a series of types. Types are the different distinguishable features of agents or strategies. From these types, successful strategies or agents emerge to help attain an agent's goals.[51] An example of variation in computer games is in a massively multiplayer online game (MMOG)[52] where there may be several thousand human agents. The large-scale game creates a population with large variation. Within these populations, there are many different types of agents. An online game may have agents varying in age from an adolescent to a senior citizen, may come from different economic backgrounds and different cultures and countries. Each of these types brings a different potential for success, therefore more variation results in more potential for successful strategies. Variation within the game depends not only on the number of agents but also on the types. In multiplayer games, variation may be highly significant, but without interaction, each of these strategies would be contained within each agent's own sphere and varieties in strategies may be slow to emerge.

Another critical concept of complexity is interaction. Without interaction, variation or selection has very little relevance. Agents interact if there is some type of connection or relationship between those agents or strategies. Interactions are external or internal, physical or conceptual.[53] Interactions in a system are what make it complex.[54] If agents did not interact, their strategies and the consequences of their strategies would be easily revealed. However when there is interaction, effects can be hidden and consequences may ripple throughout an interconnected system. Ultimately, these interactions help to make a system complex.

In *Harnessing complexity*, Axelrod and Cohen assert that selection is one of the processes of how complex adaptive systems change. They say selection is how strategies or agents are

---

[51] Robert Axelrod and Michael D. Cohen, *Harnessing Complexity*, 35.

[52] Andrew Rollings and Ernest Adams, *Andrew Rollings and Ernest Adams on Game Design*, (Indianapolis, New Riders, 2003), 344.

[53] Robert Axelrod and Michael D. Cohen, *Harnessing Complexity*, 73.

created or copied.[55] In computer gaming, there are both weak and strong selection methods, meaning that there may be a more aggressive replication of a single strategy over others. In some games, selection is strong, agents quickly copy the best strategies, and those strategies spread rapidly throughout the population, while in other games, the selection process may be weak and strategies take longer to adapt. In computer gaming, the results of these different selection methods are readily observable. In some online gaming communities, strategies may adapt and change within days, while in other games strategies may slowly adapt over months. In each of these cases, it depends on the type of game, the duration of a game, the variation among the players and the number and strength of interaction and feedback.

Feedback is what informs agents of what strategies appear to be successful. If the feedback of an unsuccessful strategy appears successful then a population of agents may select unsuccessful strategies. Therefore, it is important to understand how feedback affects the selection of agents and strategies. In some computer games, a player's character may receive damage or die when attacked. This information may provide timely feedback and a player may learn to adjust their strategies. In other computer games, players may receive small incremental changes to their character or virtual city over a long period. Further, changes may occur after many other variables have been changed. The slow and uncertain feedback may result in slow and ineffective adaptation as a player tries to identify the cause for change. In either case, feedback provides a mechanism to inform the player about the success of their strategy. Computer games could be used in a training application to show the effects of feedback because the instructor may pause or accelerate time so a student can see the effects of their decisions and

---

[54] Michael Waldrop, *Complexity: The Emerging Science at the Edge of Order and Chaos* (New York, NY: Simon and Schuster, 1992), 11.

[55] Robert Axelrod and Michael D. Cohen, *Harnessing Complexity*, 7.

reflect on them. This feedback could help students learn about the consequences of their decisions.[56]

The world that we live in is increasing in complexity. This is because the interactions among the planet's inhabitants are increasing, as globalization, information and communications technologies lead to greater interconnectivity over faster time scales. These interactions increase the possibility and rate of selection of strategies by agents in the world we live in. There are many similarities when comparing complex situations within computer gaming to complex situations in the real world. The virtual worlds created by many computer games contain the critical aspects that make an environment complex, such as agents who are seeking through strategy to accomplish a goal. During the accomplishment of an agent's goal, players interact with other agents that introduce other strategies within the system. These other agents create a variety of strategies that allow for the selection of successful strategies based on the objectives of the game. It is the combination of all these events that make computer games excellent examples of complex environments that can be harnessed to educate and train leaders about complexity and complex decision-making.

Computer Games to Educate our Leaders

The U.S. Army can educate and train its leaders to be better complex decision makers by using readily available and inexpensive COTS computer games. By doing this they can immerse them in dynamic complex adaptive environments where they have opportunities to make decisions that have complex consequences. COTS computer games can provide many different opportunities to educate and train U.S. Army leaders in complex adaptive environment through hands on, easily modifiable examples and situations that are complex in nature.

---

[56] Hassan Qudrat-Ullah, Michael J. Spector, and Paal Davidsen, *Complex Decision Making*, 1.

Computer Games

Prominent computer game designer Chris Crawford defines a game as "an interactive, goal oriented activity, active agents play against, which any player could interfere one another...."[57] Although other definitions exist, this definition is preferred because it provides an explicit link to complexity. It captures three important parts of complexity: interaction, strategy to accomplish a goal and agents. As discussed earlier in the section on complexity, complexity is dependent upon interaction among agents, and the level of complexity grows as the interaction among goal-directed agents increases. Computer games are the sub-set of games where the game state and rules of the game are stored and implemented electronically, and human agents can interact with other agents via the user interface of a computer.[58]

Capturing the importance of "what is a computer game?" is essential in understanding their value as an education tool. Chris Crawford breaks the requirements of games down further. Understanding "what a game is?" helps eliminate confusion in the discussion on gaming. Crawford makes clear that a computer game is not just a form of entertainment, a toy, or a competition.[59] Computer games must have agents that interact and compete to achieve a goal. By this definition and description, computer games contain critical components of complexity. Computer games provide opportunities for players to immerse themselves in a virtual world filled with the social structures and complexities of the real world.[60] Within these complex environments, players make decisions that can ripple through the entire system. Over time, players can learn which strategies are most effective to beat their opponents and ones that are not.

[57] Chris Crawford, *Chris Crawford on Game Design*, (Indianapolis, Ind: New Riders, 2003), 8.

[58] Alex Ryan, e-mail message to author, August 18, 2008.

[59] Chris Crawford, *The Art of Computer Game Design*, (McGraw-Hill. Osborne Media. 1986), 5.

[60] D.W. Shaffer, K. R. Squire, R. Halverson, and J. P. Gee. 2005, "Video Games and the Future of Learning". *PHI DELTA KAPPAN.* 87, no. 2: 107.

Computer games provide unique opportunities to place a player in a non-threatening environment where they can learn the effects and consequences of making decisions.[61]

Computer Games, Learning and Knowledge

Computer games contain great examples of complexity and provide an interactive environment for making complex decisions. This monograph takes a constructivist approach to learning, stating that computer games provide an environment in which a leader can be immersed to experience complexity. By being immersed in the complex environment, and interacting with other agents in competition a leader can learn to be mentally agile, use sound judgment, innovate and increase their domain knowledge.

Jean Piaget is one of the pioneers of constructivist learning theory, Piaget states that knowledge is constructed and learning occurs through an individual's experience with the world.[62] Another educational theorist is Seymour Papert. Papert is also a constructivist and adds that an individual builds knowledge through interaction with the world.[63] These theories are different from other approaches that assert that learning occurs through just doing (behaviorism) or is innate and is waiting to be untapped (inneism or rationalism).[64] Both of these constructivist theorists agree that learning should focus to create an environment in which the individual can experience and interact with their environment. Through these experiences and interactions, the individual learns by constructing knowledge based on how that individual experiences and interacts with the environment. Computer games provide an opportunity for our leaders to

---

[61] Hassan Qudrat-Ullah, Michael J. Spector, and Paal Davidsen, *Complex Decision Making*, 1.

[62] Edith Ackerman, *Piaget's Constructivism, Papert's Constructionism: What's the Difference?* http://learning.media.MIT.edu (accessed 25 September 2008), 3.

[63] Ibid., 4.

[64] Jean A. Rondal, "Another Way From the Behavior Onto the Brain and the Genes," *Friulian Journal of Schience*, no. 5, 2004: 85.

immerse themselves into a world in which they can experience complexity and then interact with that world to make complex decisions.

Another part of using COTS computer games to educate and train U.S. Army leaders is to understand the type of environment in which a leader is immersed and how they will interact. Understanding epistemology, the study of knowledge,[65] is important when constructing a learning environment for U.S. Army leaders. If leaders construct knowledge based upon their experiences and interactions with their environments, as Piaget and Papert suggest, then the environments used for learning must be epistemic or focus upon the knowledge that the U.S. Army wishes for its leaders to construct. As discussed above, the U.S. Army wants its leaders to be mentally agile, use sound judgment, innovate and have domain knowledge.[66] Therefore, the U.S. Army should create that type of learning environment for its leaders. For example, some COTS computer games focus on building a city, a zoo, a golf course or a train empire. In these games, a player must learn how to organize and manage their resources to attract people or customers. In the process of playing the game, the player must learn to think like a city planner or manager. Through interaction, the player thinks about the different variables, their interconnectedness and then changes the variables to learn about the game or system. Along the way, the player learns about how the system works and then creates solutions (innovation). A leader without a management background learns through this interaction to think like a planner or manager and increase their domain knowledge within that field. Playing the game with the proper instruction, the player may learn to use language and methods that actual planners and managers use. For example, David Shaffer in his book *How Computer Games Help Children Learn* discusses a study conducted by Gina Svarovsky at the University of Wisconsin. In this study, Svarovsky

---

[65] David Williamson Shaffer, *How Computer Games Help Children Learn* (Palgrave MacMillan, 2006), 9.

[66] U.S. Department of the Army, *Army Leadership*, 6-1.

used a program called *Digital Zoo*. This game allowed its players to create stick figure animals within a virtual environment. Unlike a normal two-dimensional drawing, the figures once created could be subjected to real gravity and put into motion. How the students create or design the animals is important to how the animals performed once put into motion. After the study, the instructor gave problems based on scientific principles to the students. According to the study, students were five time more likely to use scientific justifications to solve the problem after playing the game than before.[67] The important aspect of this study is the students learned both terminology and the application. The students designed their own creatures and through creating, learned about scientific principles. Though the students where not scientists or engineers they learned to think and talk like scientists and engineers through the gaming exercise.

Another important aspect of creating a learning environment using computer games is context and reflection. Playing a computer game without context and reflection has limited value. Shaffer states "any simulation for learning needs to be set in context if you want someone using it to develop professional vision about what is being simulated."[68] Instructors need to build the context for the game and provide the guidance during the game. This context helps ensure the student constructs the knowledge intended by the practicum. If the goal is to create a professional military leader, then the environment created should be similar to an environment that a real military leader would experience. In addition, instructors should encourage reflection throughout the game or what Donald Schön refers to as reflection-in-action.[69] Reflection-in-action occurs when a player observes that there is a difference between his expectation and his reality and therefore tries something new based on his understanding of how things work, "reflection gives

---

[67] David Williamson Shaffer, *How Computer Games Help Children Learn*, 55.

[68] Ibid., 68.

[69] Donald A. Schön, *Educating the Reflective Practitioner* (San Francisco: Jossey-Bass, 1988), 28.

rise to on-the-spot experiment."[70] This is different from reflection-on-action, which occurs after the action has occurred.[71] Shaffer points out that thinking and doing (reflection-in-action) leads to innovation.[72] True professionals do not get stuck when confronted with a complex unfamiliar situation or use the same method for each and every problem.[73] True professionals use mental agility to adjust to the situation and innovate solutions to manage problems.[74] Shaffer states, "Epistemic games deliberately identify and copy the action and reflection that develop the skills, knowledge, and epistemology of some group of true professionals. An epistemic game copies the way professionals in training learn to find innovative solutions to complex problems by systematically getting stuck and unstuck with the help of peers and mentors."[75]

As Shaffer discusses above, the final part of a learning environment using computer games to educate U.S. Army leaders requires mentorship. Mentors who facilitate learning help to create the right environment by asking questions. Those questions, as Shaffer points out, ask about a player's strategy and tactics in a game, why they choose those strategies, and what may be other courses of action.[76] Mentors also encourage risk taking and experimentation,[77]which is a major benefit of using games for educating because computer games do not have real consequences and players may learn the effects of taking risk. Dörner also discusses the importance of mentors in *Logic of Failure*, Dörner when talking about the use of simulated scenarios to teach about complex situations says that instructors should use experts to observe

[70] Donald A. Schön, *Educating the Reflective Practitioner* (San Francisco: Jossey-Bass, 1988), 28.

[71] Ibid., 26.

[72] David Williamson Shaffer, *How Computer Games Help Children Learn*, 97.

[73] Ibid., 101.

[74] Ibid., 102.

[75] Ibid.

[76] Ibid., 103.

[77] Ibid.

participants. These experts point out cognitive errors and facilitate the learning process.[78] Mentors facilitate learning by discussing cognitive lessons and ensuring the game focuses on the epistemology or learning about a profession. Mentors use the correct language and concepts about the game scenario, which help the players understand the context and how this game applies to a professional setting.

The Right Game for the Right Training Environment

There are many different genres of computer games. Popular computer games genres include first person shooters (FPS), real-time strategy (RTS), turn based strategy, (TBS), and business or economic simulation games.[79] Each of these games creates a different environment for learning. These environments offer the players many different game dynamics with which to interact. Understanding the differences between the various types of computer games and the learning environments they create is important when building a training program to educate complexity and complex decision-making. Like any training program, leaders should identify learning objectives and then select a game that best meets the requirements of those learning objectives. This section discusses a few different types of games and the learning environment that each game creates.

A FPS computer game is played in the first person point of view.[80] In these games, the player is a single agent who is looking through the eyes of the computer character. In a FPS game, the player uses a keyboard and mouse to navigate and interact with the environment. FPS games may have many goals and agents. Some of those agents can be real people and other agents can be computer artificial intelligent (AI) players. In some FPS computer games, agents

---

[78] Dietrich Dörner, *The Logic of Failure* (New York, NY: Metropolitan Books, 1996), 197.

[79] Andrew Rollings and Ernest Adams. *Andrew Rollings and Ernest Adams on Game Design.* (Indianapolis, 2003), 288.

act alone to accomplish individual goals while competing against other human players. In other FPS games, agents are on teams with other human and/or AI players. Together they try to accomplish a goal or objective. Like many games, players may play through a network within the same room or be thousands of miles apart playing via the internet. In these games, players can interact physically (within the same room), virtually (through chat) and/or voice by using voice over internet protocol (VOIP) programs.

One such FPS game is *Half Life Counter Strike: Source*.[81] Elements of complexity within this game include but are not limited to the agents, their strategies, selection and player interaction. In *Counter Strike*, strategies include selecting weaponry, attack locations, and maneuvers. Players may use different types of weaponry and, by doing so, players learn to adjust their use of weaponry based on differing circumstances. Another aspect is selecting the best locations to conduct attacks. Players may identify likely avenues of approach and wait in ambush. In other instances, players learn to use terrain and structures in which to hide as they "snipe" other players from a distance. Additionally, players learn to use maneuver and cover fire to attack and beat opponents. Players learn how to use terrain to surprise and flank opponents and to quickly maneuver into position so as to eliminate opponents within the game. Finally, players interact by being on the same team or through chat/VOIP, while opposing forces interact through game play on the scenario map.

As an education tool for military personnel, *Counter Strike* provides an environment where players use actual military terminology and tactics. In addition, when many different players play the game over time, an instructor may discuss examples of strategy and the evolution of strategy among the playing population. In addition, leaders may practice rational, intuitive,

---

[80] Andrew Rollings and Ernest Adams. *Andrew Rollings and Ernest Adams on Game Design.* (Indianapolis, 2003), 288.

[81] *Counter Strike: Source*, (PC Version) Valve Software (Vivendi Universal, 2000).

and complex decision-making. For example, an instructor can create an environment where a team develops a plan and then executes that plan, all acting together against a human opposing force (OPFOR); once the game begins a player's decision-making transitions from a rational model to an intuitive model. In this situation, the FPS *Counter Strike* is a useful training and education tool for complexity and decision-making.

Another genre of computer games is RTS games.[82] In most RTS games, players seize resources or resource points to construct buildings and produce forces.[83] Each building allows players to produce different forces such as armor, infantry or artillery. Players use these forces to attack their enemies. These events unfold in real time as all players simultaneously compete on a map. The object of the game can vary from the seizure of resources, to the protection of strategic points, to the capture of flags and/or to annihilate their enemies. Most RTS games may be played cooperatively or head to head. Additionally, players can play networked within the same room or play over the internet, and communicate using a VOIP program.

One example of an RTS game is *Company of Heroes.*[84] Specific strategies in *Company of Heroes* include rapid attacks, massing of forces, containment of the enemy, or seizure of resources. A popular technique in RTS games is to rapidly build a force and quickly attack the enemy maintaining these attacks until victorious. Another strategy is to build a massive combined arms force that moves across the map to annihilate smaller forces and then attack the enemy's bases. An effective strategy is to use terrain and forces to contain the enemy in a certain portion of the map. Finally, but not exclusively, players may try to seize and hold resources that exhaust the opponent who is unable to build without essential resources. Of course, these are

---

[82] Bruce Geryk. "A History of Real-Time Strategy Games". GameSpot. (Accessed 28 September 2008).

[83] Andrew Rollings and Ernest Adams. *Andrew Rollings and Ernest Adams on Game Design.* (Indianapolis, 2003), 233.

[84]*Company of Heroes*, (PC Version) Relic (THQ, 2006).

only a few of the popular strategies used by players, with hybrid strategies often very effective. Another aspect of *Company of Heroes* is that players may choose a specialization. In doing so, players select the type of unit capability (e.g. armor, infantry, or artillery) they wish to utilize in the game. *Company of Heroes* places players in the commander's seat, as they determine which strategy to employ while they play against an opponent in real time.

The U.S. Army could use the RTS game, *Company of Heroes*, to educate and train complexity in several ways. First, the number and variety of artifacts,[85] players use in the game, allows for the development of many different strategies. As many as eight players may play simultaneously in this game. The combination of different strategies and players that interact to seize resources, apply resources to build forces, and then develop strategies for victory helps to create a very complex environment. The complexity found in this military setting creates an interactive learning experience for U.S. Army leaders. *Company of Heroes* as an education tool provides a virtual world full of strategy, interaction, and selection while players make a wide range of decisions. An instructor using *Company of Heroes* could create an environment focused purely on understanding complexity, decision-making, or both. *Company of Heroes* as a military-based RTS game provides an environment where leaders use military terminology and techniques lending to a constructive learning experience.

Turn-based strategy (TBS) games are similar to real-time strategy games but instead, players take discrete turns, during which other agents cannot act. Actually, TBS games have a wide variety of styles of game play, many play just like a board game but use a computer to manage the game. Like a board game, TBS games can consist of simple objectives with few players and pieces or consist of many players who manage many game pieces. TBS games can also take a long time to play. The time a TBS game takes to play is based on the number of players and how much time players take between turns. If players are allowed unlimited time per turn, some TBS games can take weeks or months to play. If there is an established time limit, games can be completed in only a few hours or even minutes.

Strategy for TBS games will vary based on the type of game. A TBS game used by the School of Advanced Military Studies (SAMS) is *The Operational Art of War III*.[86] In this game, players manage massive armies reenacting historic campaigns. Depending how many players participate, one person could play the entire force against the computer or, with more players, the forces may be divided among the many players. The number of players creates the complexity within this game. If leaders role-play different army commanders, as they do at SAMS, then the game can become very complex. Each of the army commanders develops their strategy for their portion of the war. These commanders then interact among each other identifying objectives and targets while they request help or forces from other commanders. For example, in one SAMS exercise, SAMS students played the German High Command, re-enacting Operation Barbarossa versus a computer-controlled Soviet Union.[87] *The Operational Art of War III* game allows time for players to think through possible current moves in conjunction with opponent current and future moves. This extra time to make a decision allows players to optimize their resources and plan as they try to predict their opponent's reactions and subsequent moves. The unlimited time between turns provides an opportunity for players, if they wish, to use a rational or analytical decision-making model.

The Operational Art of War III may serve as an educational tool for complexity and complex decision-making for several reasons. In *The Operational Art of War III*, players may participate in a complex environment where they must interact with other agents (commanders) to

---

[85] Robert Axelrod and Michael D. Cohen *Harnessing Complexity*, 6.

[86] *The Operational Art of War III*, (PC Version) Norm Koger (Matrix Games, 2006).

[87] *The Operational Art of War III* has an editor that allows individuals to change scenarios or create new ones, which may then be played within the game. Games with editors make for good education tools because of the ability to change scenarios.

pursue their goals. Within the game itself and among the commanders, a player must weigh the importance of relationships between their peers, the best use of units, and the pursuance of an objective within the game. All players may pursue many strategies and use many different units. In each turn, a player uses many units to battle, to move, and/or to secure objectives. During these interactions, it is difficult to establish cause and effect and in some cases the effects of decisions made early within the game may not be known until later turns. *The Operational Art of War III* is great for making complex decisions because it encompasses the elements of Strohschneider's definition of a complex decision such as complexity, a dynamic environment, the effects of decisions in the game are ambiguous and players must make a large number of decisions each turn. All these elements combine to create an environment, which allows players to make complex decisions.

Another genre of computer games is business or economic simulations games.[88] These games create an environment similar to a real business or economic situation. There is a wide variety of ways in which a player interacts with these simulations. Some games may have graphics with moving agents and artifacts, while others may be a series of screens with numbers, charts and graphs. In each case, players must balance a series of economic variables to create a successful business.[89] The challenge in economic simulation games is the complex interactions between different variables. For example, one variable is required to produce another and the production of a variables may create intended and unintended consequence somewhere else within the game. The interdependence and uncertainty help to create a complex environment.

---

[88] Andrew Rollings and Ernest Adams. *Andrew Rollings and Ernest Adams on Game Design.* (Indianapolis, 2003), 234.

[89] Ibid., 235.

One such business simulation game is Sid Meier's *Railroads!*,[90] which is a game based on the railway industry. In this game, players compete to make money by linking resources, industry, and people through a rail system. Players of this game, may employ many different strategies and techniques to win the game. They must find raw materials such as grapes and then find an industry to convert those raw materials into a product such as wine. These products are then transported via railroad to a town that consumes the wine. As the towns make more money, the town grows larger and consumes more goods. The money made from the consumption of the goods goes to the rail tycoon who can then build more railroads or upgrade the rail system. Players in *Railroads!*, must determine which resources are worth transporting over long distances and what types of trains and rail they should build. *Railroads!* is also a RTS game; therefore players compete in real time against one another to seize resources, build products, and build market share. The many variables, players, and strategies in competition within a real time situation lend to create a very complex and dynamic learning experience.

The computer game *Railroads!* may be used as an education tool for multiple reasons. In the business simulation game *Railroads!*, players compete against each other in a dynamic business market. Players within the game make decisions on what types of commodities and raw materials they wish to deal in. Then players work with industry to produce a product that is then sold to consumers within each city. Players learn about supply chain and business ideas. In addition to learning business concepts, players learn about the importance of how to allocate resources to make railroads and purchase trains. Another aspect of the game is players must also manage time schedules for each of their train routes. By managing their routes, players can maximize the capacity of their trains and reduce expenditures. Though *Railroads!* is not a military type of game, this game can be used to teach business concepts to military leaders. In this game, U.S. Army leaders may make complex decisions as they learn about supply and

---

[90] *Railroads!*, (PC Version) Firaxis Games (2K Games, 2006).

demand, supply chain management, production management, income, and expenditures. This learning opportunity combined with a fast-paced dynamic environment creates an interactive learning experience where leaders can learn about complexity and make complex decisions.

In summary, there are different types of computer games, including first person shooters, real-time strategy, turn-based strategy and business or economic simulation computer games. Each type of computer game creates a different type of gaming environment demanding different strategies and interactions. An instructor may take advantage of the differences between the gaming environments to create an educational experience to teach and train different aspects of complexity and complex decision-making, all with real world applicability and implications.

Training Decision-Making in a Complex World

Thus far, this paper has argued that the U.S. Army can train and educate its leaders to be better decision makers within a complex environment by using COTS computer games. The next part of this paper reviews how the U.S. Army trains its leaders. Specifically, it discusses "how does the U.S. Army train its leaders in decision-making?" By understanding how the U.S. Army currently trains its leaders for decision-making, this paper can then make recommendations as to how COTS computer games should be integrated to teach complexity and complex decision-making. The last part of this section presents a method for using COTS computer games to train and educate its leaders in complexity and complex decision-making.

Figure 3.1: U.S. Army Training and Leader Development Model

The U.S. Army uses the Army Training and Leader Development model to train and develop its leaders (Figure 3.1).[91] This model focuses on three core domains to train and educate its leaders. Those three domains are operational, institutional, and self-development.[92] Through these domains, the U.S Army hopes to prepare its Soldiers and leaders to operate in a world of complexity. The operational training domain focuses on training Soldiers through operational deployment, home station training, and training at the U.S. Army's combat training centers (CTC).[93] The institutional training domain educates and trains U.S. Army leaders through unit and joint training schools and advanced education opportunities.[94] The self-development training

---

[91] U.S. Department of the Army, *Training the Force: Field Manual 7-0,* (Washington, DC, 2002), 1-6.

[92] U.S. Department of the Army, *Training the Force,* 1-5.

[93] Ibid., 1-9.

[94] Ibid., 1-7.

domain is the formal and informal education that fills gaps that may exist between the operational and institutional training domain.[95]

Where does training complexity and complex decision-making fall within the Army Training and Leader Development Model? Training manual FM 7-0 states that the institutional training domain prepares and educates our leaders to be critical thinkers who are able to deal with uncertainty. The manual goes on to state that the institutional domain prepares leaders to be mentally agile.[96] Many different techniques can be used to increase U.S. Army leaders' mental agility and ability to make complex decisions. The U.S. Army uses mostly classroom instruction, case studies, and practical exercises to educate and train its leaders to become flexible, innovative and mentally agile critical thinkers.[97] These education techniques have their merits but COTS computer games provide a complementary way to train Army leaders within all three training domains.

COTS computer games provide an excellent training and education tool that can be used in the institutional training domain. The U.S. Army has one of the best institutional training programs in the world. Thousands of new recruits, young officers, and seasoned sergeants and officers attend many of the U.S. Army's premier institutions each year. This domain uses a variety of methods to include classroom instruction, field instruction, range training and simulation training to educate and train our future leaders. Another inexpensive opportunity would be the addition of COTS computer games as an education tool to instruct complexity and complex decision-making. COTS computer games bring an entertaining and familiar training tool to U.S. Army schoolhouses. Many young Soldiers may already be familiar with several of the computer game genres discussed in this monograph. Additionally, some COTS games have

---

[95] U.S. Department of the Army, *Training the Force*, 1-11.

[96] Ibid., 1-7.

[97] Ibid., 1-8.

interactive tutorials, therefore little time is required to learn the mechanics of the game and maximum time may be spent learning about complexity and complex decision-making.

The operational training domain could also benefit by using COTS computer games to train and educate complexity and complex decision-making. The operational training domain relies on operational deployments and field training exercises to prepare Soldiers to operate on today's battlefield.[98] In many cases, commander's struggle to find time and resources to provide realistic and battle-focused training for their Soldiers COTS computer games can make available to commanders another facet of training that can be done within the unit. Though operational experience is priceless, operational deployments and exercises entail risk and are one-time events. There are no reset buttons during real combat operations. Therefore, in real operations, leaders get one chance to make decisions and then must live with the consequences of those decisions. While COTS computer games can never replace the value of real operational experience, they do provide an inexpensive, repetitious, and reviewable situation where leaders can make decisions and then receive feedback from those decisions.

How can a commander use COTS to train their leaders? A method for commanders to utilize COTS computer games at the unit level is to setup a computer exercise area. This area would consist of about ten computers, which would facilitate squad training (nine squad members plus an instructor). Leaders could create training scenarios to achieve specific learning objectives using the COTS computer games. For example, leaders could use a FPS in order to train teamwork and communication skills while they evaluate small unit leaders as they make decisions. This computer area could serve a dual purpose when not in use for training. It could provide a computer lab for soldiers to use for work, professional development, or morale, welfare, and recreation (MWR) uses.

---

[98] U.S. Department of the Army, *Training the Force*, 1-5.

Finally, COTS computer games can be used in the third training domain, self-development. This domain relies on leaders to continually train and educate themselves, much of which is typically done at home or on personal time.[99] COTS computer games provide a training technique they may use at work, school, or at home. Therefore, the U.S. Army could develop a comprehensive training programs and lessons using COTS computer games that are tailored to be used at Army schools, at their units, and at home. These lessons may be posted online and leaders, while at home, could access the site, read about a subject or watch a video, and then play a game under the context of the learning objectives established by the U.S. Army. After the game, the leader could participate in a survey, quiz or discussion that helps the leader reflect on the learning objectives.

An inexpensive COTS computer game the U.S. Army could use to begin a training program would be the U.S. Army's *America's Army*.[100] *America's Army* is a FPS game that is free for download from the U.S. Army's *americasarmy.com* website. This game is similar to many FPS games such as *Counter Strike* and *Unreal Tournament*[101]. As discussed above in the FPS section, instructors could create lessons that focus and discuss the strategies and interactions for each player and team. Over time, leaders could discuss how strategies adapt and how leaders playing the game must make complex decisions that include understanding the different team member's abilities, situations and technical capabilities. Through the game, leaders could learn to consider all the different facets of making complex decisions, practice making decisions, and then receive feedback on those decisions. In addition, *America's Army* is a computer game based purposely on the U.S. Army. *America's Army* creates an epistemic learning environment where leaders learn and practice the language, concepts and techniques used by professional Soldiers.

---

[99] U.S. Department of the Army, *Training the Force*, 1-12.

[100] *America's Army* (PC Version) U.S. Army (U.S. Army, 2002).

[101] *Unreal Tournament* (PC Version) Epic Games (GT Interactive, 1999).

This epistemic learning environment creates a constructive military experience where they can interact and perform while making complex decisions.

Unlike other training tools, COTS computer games can be used across all three training domains. Leaders can use COTS computer games to train new recruits at the institutional level, at their unit during operational training, or as self-development at home. Soldiers can use COTS computer games to practice decision-making and learn about the components of complexity while sitting in their barracks, in a company training area, or home office. As discussed, the U.S. Army can develop training plans that encompass all three training domains using a single software platform to create a comprehensive training program that may be used throughout a Soldier's career.

Benefits of Using COTS Computer Games

There are many benefits for the U.S. Army to use COTS compute games to educate its leaders in complexity and complex decision-making. One benefit of COTS computer games are that COTS computer games place leaders in a risk free environment to make decisions. Another benefit is that COTS computer games subject leaders to complex situations and provide quick feedback allowing leaders to learn from their experiences. Finally, COTS computer games are a cost effective platform for training leaders in complexity and complex decision-making.

COTS computer games place leaders in a virtual world that allows leaders to make complex decisions without suffering the consequences of making mistakes from those decisions.[102] Unlike a real operating environment, where decisions can have dire consequences, a computer game environment is a nonthreatening environment where leaders can more freely make decisions. This opportunity allows leaders to take risks not normally possible in a real world situation. These risks may help leaders understand the relationships between their

---

[102] David Williamson Shaffer, *How Computer Games Help Children Learn*, 68.

decisions and the consequences of those decisions. In computer games, leaders may receive feedback from those consequences almost immediately because of the compression of time and space.[103] By receiving feedback, leaders can learn the importance of how strategies must adapt and change in response to environmental changes - concepts important to understanding complexity.

Intuitive decision-making or natural decision-making relies heavily on a leader's prior experience.[104] As discussed earlier, the U.S. Army wants its leaders to use sound judgment when making decisions. For a leader to make sound decisions a leader must be able to identify the type of problem they confront and all the different variables affecting that problem. COTS computer games provide a venue to repetitively subject a leader to a complex situation and receive feedback. By repetitively subjecting a leader to a complex adaptive environment they gain the experience that helps them build mental simulations for complex situations. Thus, leaders are better equipped to accurately identify the problem they face and the relevant variables impacting that situation. Therefore, U.S. Army educators can use computer games to provide an environment in which leaders can use both analytical and intuitive decision making to test and experiment ideas. Furthermore, this training may help leaders learn to develop courses of action and to innovate solution sets to a problem. For instance, leaders may need to act quickly under a time constraint, while in other situations, leaders may have plenty of time to evaluate the problem. In either case, leaders must rely on their experiences and domain knowledge to develop and implement solution sets or make a complex decision. A benefit of COTS computer games is they place leaders in situations where they can practice making complex decisions and develop decision-making skills.

---

[103] Hassan Qudrat-Ullah, Michael J. Spector, and Paal Davidsen, *Complex Decision Making*, 1.

[104] U.S. Department of the Army, *Mission Command*, 2-4.

Another benefit of using COTS computer games to educate complexity and complex decision-making is the immediate feedback a player receives from computer games. Decisions in a real situation may conceal second and third order consequences for a long period of time (months or even years). In a computer game, the push of a button or the click of a mouse speeds up or slows down time. Consequences of decisions are revealed almost immediately. Feedback is critical in adaptation of strategies in that it reveals successful and unsuccessful actions. Leaders can then use measures of success to identify successful actions that lead to the adaptation of strategy. By focusing on these instances, junior leaders can learn how to adapt strategies for future actions. By learning the role of feedback in adapting strategy within a game, leaders can learn how to identify feedback loops within a real situation. By learning how to identify feedback loops, leaders may become better complex decision makers.

The U.S. Army states in its leadership manual that it wants leaders, who are mentally agile, innovative, flexible, adaptive and able to deal with uncertainty.[105] In addition, in the U.S. Army training manual, it specifically outlines how it wants to train U.S. Army leaders to be mentally agile, innovative, flexible, adaptive, and able to deal with uncertainty.[106] As key characteristics of leadership, how can the U.S. Army then train its leaders to maximize their potential to encompass these critical decision-making attributes? By studying and learning how to harness complexity, a leader can make relevant decisions within complex environments. In addition to understanding how to take advantage of complexity, leaders need to practice decision-making. COTS computer games provide a great environment for leaders to practice decision-making and to learn about the consequences of making decisions within a complex environment.

Each of the U.S. Army training domains require significant resources and time to accomplish the learning and training objectives. In addition, these training domains have large

---

[105] U.S. Department of the Army, *Army Leadership*, 2-4.

[106] U.S. Department of the Army, *Training the Force: Field Manual 7-0*, 1-7.

personnel requirements. All of these factors combine to create a rigid training environment. In this environment, leaders must adhere to strict training schedules, fall within budgetary guidelines, produce and update tedious and detailed training plans; all of which consume valuable time and resources. COTS computer games do not require as much personnel to establish a training program. Such a training program may require but one individual to setup, operate and monitor training. Another benefit is very little training is required to learn the mechanics of the game. Most COTS computer games provide a built-in tutorial that guide players through the mechanics of each game. For instance, if a training program is developed around the U.S. Army's *America's Army* computer games then little funds are required. With the right training plan and focus, COTS computer games use minimal resources, time and personnel to establish and maintain. A COTS computer game training and education program can help ensure today's leaders have an opportunity to learn and practice making decisions within a complex adaptive environment.

Conclusion

Make-believe has always been an important way to prepare ourselves for the real thing. We should use this method in a focused manner. We now have far better tools for this purpose than we ever had before. We should take an advantage of them.

Is this a frivolous idea? Playing games in dead earnest? Anyone who thinks play is nothing but play and dead earnest nothing but dead earnest hasn't understood either one.[107]

Deitrich Dörner

As Dörner points out above, playing games in a focused or "dead earnest" manner is a great way to prepare our leaders for real world situations. Dörner then rightly points out that there are great tools available to prepare our leaders for those situations. As argued in this monograph, COTS computer games can serve in this role. Senior U.S. Army leaders have an opportunity to restructure their thoughts on what a computer game is and how a computer game can be used. This monograph contributes to reshaping Army thinking by showing that gaming does indeed have leader development applications by helping Army leaders practice decision-making in a complex and uncertain world.

The U.S. Army has an opportunity that other organizations like Duke University's Anesthesiology Department or even our adversaries are now benefiting from. There is an opportunity to wield the benefits of COTS computer games to educate and train leaders within the U.S. Army. COTS computer games provide a method for educating our leaders about complexity and then provide a virtual complex environment, in which leaders can be immersed in order to practice making complex decisions.

---

[107] Dietrich Dörner, *The Logic of Failure*, 199.

The U.S. Army demands leaders with the intellectual capacity to make sound, relevant decisions. There may be no single way to improve intellectual capacity but according to Papert and Piaget, individuals construct knowledge through experience and interaction with their environment.[108] This monograph argues that computer games provide a complex environment that U.S. Army leaders can experience and interact with, which provides an epistemic learning experience. As discussed, computer games offer many situations where a player can exercise mental agility by recognizing rival strategies and then formulate counters to those strategies. A player may use their knowledge of the game to select a pre-existing strategy or innovate by formulating a new one. Finally, the player uses their judgment and makes a decision. From the decision comes action, as the player begins movement either on their turn or in real time. Whatever the case may be, the player is interacting and practicing in a complex environment. The leader is not just practicing how to play the game, per se, but practicing to identify problems, formulate solutions and implement those solutions. Computer games provide the ultimate practice field; a field with limitless arrangements and opportunities to practice complex decision-making.

Finally, COTS computer games can be used within all three of the U.S. Army training domains. Unlike other training systems, COTS computer games are inexpensive and readily available. These aspects allow COTS computer games to be utilized in a classroom, in a company training room, or at home. The U.S. Army may develop training programs that allow Soldiers the opportunity to use a single platform from Advanced Individual Training (AIT), through the operational training domain at their unit, to self-development at home. Furthermore, COTS computer games offer the opportunity to educate and train Soldiers throughout their careers, from initial entry in the service to a seasoned military leader.

---

[108] Edith Ackerman, *Piaget's Constructivism, Papert's Constructivism: What's the* Difference? (Massachusetts Institute Of Technology), 3.

In conclusion, this monograph has shown that the U.S. Army has an opportunity to utilize COTS computer games as a constructive learning tool to educate and train its leaders in complexity and complex decision-making. Although understanding complexity and being able to make relevant decisions are important aspects of leadership, other opportunities and benefits of COTS computer games exist. In addition to teaching complexity and complex decision-making, further research can be conducted that discusses the benefits of using COTS games to practice communication, teamwork, team-building, problem solving, and possibly instruct and practice operational tactics, techniques and procedures (TTP). Given a chance, COTS computer games may play a future role in U.S. Army training and education.

BIBLIOGRAPHY

America's Army, (PC Version) Developer: U.S. Army / Publisher: U.S. Army, 2002.

Ackermann, Edith. *Piaget's Constructivism, Papert's Constructionism: What's the difference?* http://learning.media.mit.edu/content/publications/EA.Piaget%20_%20Papert.pdf (accessed 29 September 2008).

Ashby, W. Ross. *An Introduction to Cybernetics*. New York: J. Wiley, 1956.

Aldrich, Clark. *Simulations and the Future of Learning An Innovative (and Perhaps Revolutionary) Approach to E-Learning*. San Francisco: Pfeiffer, 2004.

Aldrich, Clark. *Learning by Doing: A Comprehensive Guide to Simulations, Computer Games, and Pedagogy in E-Learning and Other Educational Experiences*. San Francisco, CA: Pfeiffer, 2005.

Chatham, R. E. 2007. "Games for Training". *COMMUNICATIONS- ACM.* 50, no. 7: 36-43.

Company of Heroes, (PC Version) Developer: Relic / Publisher: THQ, 2006.

Counter Strike: Source, (PC Version) Developer: Valve Software / Publisher: Vivendi Universal, 2000.

Crawford, Chris. *The Art of Computer Game Design*. Berkeley, Calif: Osborne/McGraw-Hill, 1984.

Crawford, Chris. *Chris Crawford on Game Design*. Indianapolis, Ind: New Riders, 2003.

Davenport, Thomas H. *Thinking for a Living: How to Get Better Performance and Results from Knowledge Workers*. Boston, Mass: Harvard Business School Press, 2005.

Delwiche, A. 2006. "Massively Multiplayer Online Games (MMOs) in the New Media Classroom". *JOURNAL OF EDUCATIONAL TECHNOLOGY AND SOCIETY.* 9, no. 3: 160-172.

Dörner, Dietrich. *The Logic of Failure: Why Things Go Wrong and What We Can Do to Make Them Right*. New York: Metropolitan Books, 1996.

Dreyfus, Hubert L., Stuart E. Dreyfus, and Tom Athanasiou. *Mind Over Machine: The Power of Human Intuition and Expertise in the Era of the Computer*. New York: Free Press, 1986.

Gee, James Paul. *What Video Games Have to Teach Us About Learning and Literacy*. New York: Palgrave Macmillan, 2003.

Gee, J. 2005. What would a state of the art instructional video game look like?. *Innovate* 1 (6). http://www.innovateonline.info/index.php?view=article&id=80 (accessed October 10, 2008).

Geryk, Bruce. "A History of Real-Time Strategy Games". GameSpot. http://www.gamespot.com/gamespot/features/all/real_time/index.html. (Accessed 28 September 2008).

Glouberman, S., and Brenda Zimmerman. *Complicated and Complex Systems What Would Successful Reform of Medicare Look Like?* [Saskatoon]: Commission on the Future of Health Care in Canada, 2002.

Jong, M.S.Y., J. Shang, F.-L. Lee, and J.H.M. Lee. 2008. "Harnessing Computer Games in Education". *INTERNATIONAL JOURNAL OF DISTANCE EDUCATION TECHNOLOGIES.* 6, no. 1: 1-9.

Kelly, Anthony. *Decision Making Using Game Theory: An Introduction for Managers*. Cambridge, UK: Cambridge University Press, 2003.

Klein, Gary A. *Sources of Power How People Make Decisions*. Cambridge, Mass: MIT Press, 1999.

Ko, Seonju. 2002. "An Empirical Analysis of Children's Thinking and Learning in a Computer Game Context". *Educational Psychology*. 22, no. 2: 219-233.

Otto, Richard Forbes. *A Study of the Massively Multiplayer Online Role Playing Game Everquest: Can a Virtual Game World Be a Community? : the Voices of Netizens of Norrath*. Thesis (Ph.D)--University of Memphis, 2007.

Peck, Michael, "Constructive progress: U.S. Army embraces games — sort of," *TSJ Online.com: Training & Simulation Journal* (December 2007) http://www.tsjonline.com/story.php?F=3115940, (accessed March 2008).

Qudrat-Ullah, H., J. Michael Spector, and P. I. Davidsen. *Complex Decision Making: Theory and Practice*. Understanding complex systems. Berlin: Springer, 2008.

Quinn, Clark N., and Marcia L. Connor. *Engaging Learning: Designing E-Learning Simulation Games*. San Francisco, CA: Jossey-Bass, 2005.

Sid Meier's Railroads!, (PC Version) Developer: Firaxis / Publisher: 2k Games, 2006.

Robbins, Stephen P., and Timothy A. Judge. *Organization Behavior*. 12th ed. Upper Saddle River, New Jersey: Pearson Prentice Hall, 2007.

Rollings, Andrew, and Ernest Adams. *Andrew Rollings and Ernest Adams on Game Design*. Indianapolis, Ind: New Riders, 2003.

Rupinta, Amber, "Medical Students Using Games to Practice", WTVD-TV/DT (March 2008) http://abclocal.go.com/wtvd/story?section=news/health&id=5996041, (accessed March 2008).

Schön, Donald A. *Educating the Reflective Practitioner*. San Francisco: Jossey-Bass, 1988.

Shaffer, David Williamson. *How Computer Games Help Children Learn*. New York: Palgrave Macmillan, 2006.

Shaffer, D. W., K. R. Squire, R. Halverson, and J. P. Gee. 2005. "Video Games and the Future of Learning". *PHI DELTA KAPPAN*. 87, no. 2: 104-111.

Shaffer, D. 2005. Epistemic games. *Innovate* 1 (6). http://www.innovateonline.info/index.php?view=article&id=79 (accessed October 10, 2008).

Smed, Jouni, and Harri Hakonen. *Towards a Definition of a Computer Game*. TUCS technical report, no 553. Turku: Turku Centre for Computer Science, 2003.

Strohschneider, Stefan, *Cultural factors in complex decision-making, Online Readings in Psychology and Culture*. Center for Cross-Cultural Research, Western Washington University, Bellingham, WA, 2002.

The Operational Art of War III, (PC Version) Designer: Norm Koger / Publisher: Matrix Games, 2006.

Unreal Tournament, (PC Version) Developer: Epic Games / Publisher: GT Interactive, 1999.

U.S. Department of the Army, *Army Planning and Orders Production: Field Manual 5-0*, Authpub-FM. Washington, DC, 2006.

U.S. Department of the Army, *Army Leadership: Competent, Confident, and Agile: Field Manual 6-22*, Authpub-FM. Washington, DC, 2006.

U.S. Department of the Army, *Mission Command: Command and Control of Army Forces Field Manual 6-0*, Authpub-FM. Washington, DC, 2003.

U.S. Department of the Army, *Training the Force: Field Manual 7-0*, Authpub-FM. Washington, DC, 2002.

Waldrop, M. Mitchell. *Complexity: The Emerging Science at the Edge of Order and Chaos*. New York: Simon & Schuster, 1992.